From Global Land Grabbing for Biofuels to Acquisitions of African Water for Commercial Agriculture

David Ross Olanya

NORDISKA AFRIKAINSTITUTET, UPPSALA 2012

INDEXING TERMS:
Africa
Land acquisition
Biomass energy
Fuel
Water resources
Agricultural production
Commercial farming
Foreign investment
Property rights
Livelihood
Environmental aspects
Government policy

ISSN 0280-2171
ISBN 978-91-7106-729-6
© The author and Nordiska Afrikainstitutet 2012
Production: Byrå4
Print on demand, Lightning Source UK Ltd.

Contents

List of figures and tables

Abstract

Expansion of biofuel investment in Africa has been supported by indebted poor governments because of perceived potential benefits such as sustainable energy development, support to poor farmers, development of rural economies and reduction of greenhouse gas emissions. However, the intensity of the biofuels political economy in poor countries worsens inequality for the vulnerable poor. This is evidenced by large-scale land acquisitions in Africa for biofuel and crop production primarily for foreign consumption – food, animal feeds and energy crops. The search for land in African countries has been triggered by growing concerns over food and energy security in developed countries following the global food crisis of 2008. Moreover, these recent developments in large-scale land acquisitions in Africa are not a new phenomenon, but represent the renewal of old practices in commercial agriculture, which is either conducted through purchases or long-term leases.

In addition to biofuel expansion, this study notes that current large-scale land acquisitions in sub-Saharan Africa have been further driven by demands to access water resources for other commercial agricultural crops. The land purchases or leases automatically guarantee access to African water. This demand for water is a response to climate change: most industrialists believe that acquiring land near a main water reservoir will guarantee future agricultural potential. Few analyses have been done on the land-water access nexus. This article considers recent developments in large-scale land acquisitions in Africa in terms of water security for commercial agriculture to safeguard the production of agricultural crops with a large water footprint. Using political economy analysis, this article examines national policy on these acquisitions, the rights accorded to foreign investors and how land acquisitions undermine indigenous rights to the common resources that have been the main source of livelihood in sub-Saharan Africa.

Keywords: land acquisition, sustainability, common resources, water, water footprint

The current global 'land grab' is causing radical changes in land use and owner-ship. As pointed out by Borras and Franco (2010:2), the phrase 'land grab' aptly describes the current commercial land transactions for the production of food and biofuels in developing countries. According to reports in the media and the emerging literature, the main process driving the 'land grab' is the production of food for export by finance-rich, resource-poor countries and biofuels for export after the 2007–08 food and energy crises (Zoomers 2010:429). The term is used to describe the enclosure of commons, dispossession of peasants and indigenous people of the South of land by foreign governments and companies (Zoomers 2010:429). First used by environmental and agrarian justice movements opposed to the land transactions, the term has now been absorbed into current depoliti-cised mainstream development policy, with the push for 'win-win' arrangements and a 'code of conduct' (Borras and Franco 2010:2). It is being 'reframed' by actors, including capitalist agro-industry, to mean a golden opportunity in the name of pro-poor and ecological economic development (Borras and Franco 2010: 3). Generally, it describes large-scale, cross-border land deals or transac-tions involving transnational corporations or initiated by foreign governments (Zoomers 2010:429).

Much of the 'land grab' results from increasing demand for cheap foodcrops. Food supply problems have resulted from production bottlenecks in storage and distribution, while expansion of biofuels competes with local land uses. Host governments also generally welcome foreign investment, even though their own populations lack food. In fact, 'food-insecure' governments that used to rely on imports to feed their populations, like the Gulf States, are now seeking to outsource domestic production by buying and/or leasing vast areas of farmland abroad for food production. This explains why state-owned firms in Qatar, UAE and other Gulf States are reportedly involved in land acquisitions in Egypt, Su-dan, Ethiopia and other Africa countries (Zoomers 2010:434–5).

National governments in finance-rich, resource-poor countries are look-ing to finance-poor, resource-rich countries to help secure their own food and energy needs into the future. Control of large landholdings in other countries is needed for this purpose (Borras and Franco 2010:4). The 2007-08 boom in food prices and the subsequent period of relatively high and volatile prices reminded many import-dependent countries of their vulnerability to food in-security and prompted them to seek new opportunities to secure food. How-ever, with little empirical data about the scale of this phenomenon, opinions about the implications are divided. Some commentators see it as an opportu-nity to reverse longstanding underinvestment in agriculture that could allow land-abundant countries to gain access to better technology and more jobs for poor farmers and other rural citizens. If managed well, this trend could create the preconditions for sustaining broad-based development (Deininger *et al.* 2011).

Recipients of these investments are poor developing countries that are actively trying to attract investors because they value land deals as an opportunity to gain funds for the development of agriculture or infrastructure (Friss and Reenberg 2010:7). The global 'land grab' is to a large extent the result of the liberalisation of land markets, a policy adopted in the early 1990s that has contributed to the commoditisation of land and other natural resources (cf. Brenner and Theodore 2007, in Zoomers 2010:431). International institutions rhetorically facilitate the process of extensive land purchases in developing countries in the belief that such deals provide 'win-win' situations for investors and 'host' countries. The principal actor among these institutions is the International Finance Corporation, the private sector arm of the World Bank Group, which finances private investments in developing countries and provides advice to governments to create business-enabling environments (Daniel and Mittal 2009:6). Other players legitimising land grabs as 'win-win' include donor governments, research institutions and international governmental agencies, including the Food and Agricultural Organisation (FAO) and other UN agencies. They base their arguments on the promotion of economic development in poor countries through the provision of jobs in agriculture and other linkage industries to boost exports and through the acquisition of new technology to improve farm efficiency (Daniel and Mittal 2009:9). The International Food Policy Research Institute believes that transparency in negotiations, respect for existing rights and sharing benefits between local communities and foreign investors can provide key resources for agriculture, including the development of needed infrastructure and alternative livelihood options. However, the concept of 'food sovereignty' can only be secured by promoting small farming as a key to enhancing food production (McMichael 2010: 613). This perspective, however, has been neglected by the international development agencies: instead they support individualisation and large-scale farming for export.

Much literature exists on the expansion of biofuels and the associated 'land grab' across the world (e.g., GRAINS 2008; Cotula *et al.* 2009; von Braun and Meinzen-Dick 2009). Few studies have, however, been conducted explicitly on the political economy of biofuel development (e.g., Borras *et al.* 2010; McMichael 2010; Dauvergne and Neville 2009, cited in White and Dasqupta 2010). Even less information exists on the nexus between water and land acquisitions (Smaller and Man 2009). Political economy approaches explicitly ask at least four fundamental questions: Who owns what? Who does what? Who gets what? And what do they do with the agrarian surplus? (Bernstein 2010). There is scant information on the land-water nexus and hydro-security. This is the gap explored in this study. Since it is a preliminary study, it can be used to guide future studies.

This article reviews existing information from GRAIN, International Institute for Environment and Development, Food and Agricultural Organisation, International Land Coalition, and documentation from government agencies and other international organisations. It also uses secondary literature and other information related to biofuels. The geographical scope of the study is sub-Saharan Africa, which has abundant natural resources, including fertile tropical land, rivers and lakes, and coastal areas with humid temperatures. All these conditions are favourable for biofuel production. This study assumes that land in these areas is held under a dualistic regime, with statutory and customary ownership co-existing. Regardless, land acquisitions mostly take place without respect for the security and livelihoods of indigenous peoples and their access to common resources.

This article examines the land-water acquisition nexus, and argues that recent large-scale land acquisitions in sub-Saharan Africa are driven by concerns to secure water for commercial agriculture against climate change. Investments in farmlands in Africa are seen as a water insurance policy for posterity against increasing food shortages, declining water supplies and climate change, coupled with huge population increases. The search for land-water access in Africa is driven by the dual factors of climate change and population growth that will combine to squeeze water resources and affect food security in Gulf States, emerging and developed countries, regardless of how water-secure they may be today. At that point, water scarcity will be a potential constraint on economic development and could create social unrest, especially when dwindling resources result in higher prices and limited access for people. Using the concept of water footprint, this article further explores how water-poor, finance-rich countries depend on external water resources. In addition, the study notes that biofuel projects target African water resources, as opposed to the belief that biofuels utilise African marginal lands.

This rest of this article is divided into five sections. To initiate the discussion of the land-water access nexus, the first two sections review empirical information on biofuels and large-scale land acquisitions in Africa. The first provides a brief overview of the drivers, actors and impacts of large-scale land acquisitions in Africa. In the second section, the changing land relations among actors in large-scale land acquisitions are explored as well as the moves by proponents of a win-win approach for rural communities, host governments and investors in agricultural development in Africa. The third section debates contemporary land-water access, and considers land acquisitions for biofuel and foodcrop production to be more attractive in water-rich areas than in water-poor areas of the continent. Using the concept of water footprint, water-scarce countries such as

Gulf States that used to have high virtual-water imports have been affected by a food crisis. Foodcrops such as wheat and maize are being converted into liquid biofuels in water-rich countries. The ultimate alternative is land acquisitions in Africa to produces foodcrops.

SECTION A: The Drivers, Actors and Impacts of Large-Scale Land Acquisitions in Sub-Saharan Africa

The Drivers

Large-scale land acquisition broadly means purchases or leases of land areas ranging from 1,000 hectares to 500,000 hectares (Cotulal *et al.* 2009:3) by wealthier food-insecure nations and private investors to produce export crops (Daniel and Mittal 2009:2). The immediate short-term drivers of land acquisitions were the high food prices in 2008 and high oil prices in 2007 and 2008. These prompted private investors and banks to search for new sources of investment in the property sector (Smaller and Mann 2009:4). In general, earlier 'land grab' scholars identified three main considerations in agricultural land investments: food security; demand for biofuels; and alternative sources of energy amidst climate change (Daniel and Mittal 2009:2; Brittaine and Lutaladio 2010:4).

The food security of food-insecure nations was threatened by skyrocketing food prices in 2008 resulting from increased import bills and inflation, harsh weather conditions, poor soils and scarce land and water in many areas, combined with economic and demographic growth (Daniel and Mittal 2009:2). Also, food security concerns in investor countries have been the key driver in government-backed investments. Food supply problems and uncertainties were created by constraints in agricultural production because of limited availability of water and arable land; bottlenecks in storage and distribution; and the expansion of biofuel production. Increasing urbanisation rates and changing diets are also pushing global food demand (Cotula *et al.* 2009:4). Biofuel production has led to monoculture and affected the production of traditional crops (Matondi 2011).

With regard to biofuels, government consumption targets (in the European Union, for instance) and financial incentives have been the key driving force (Cotula *et al.* 2008:1). There are rising rates of return in agriculture. Rising commodity prices in particular make the acquisition of land for agricultural production an increasingly attractive option. The improved investment climate in several countries stemming from investment treaties and codes and land legislation has also boosted land markets in Africa (Cotula *et al.* 2009:5).

However, recently the driving forces have been broadened to include population increase and water scarcity in the Gulf States and the emerging economies of countries like China and India. Other factors include the need for rural development and export development in home countries; demand for non-food agricultural commodities; expectation of returns in the private sector; emerging carbon markets – especially biofuel projects and the long-term REDD (Reducing Emissions from Deforestation and Forest Degradation in Developing

Countries) scheme under the Kyoto climate change regime; and host country incentives (Cotula *et al.* 2009:52–8). What is important in this context is that biofuels produce fewer particulates, hydrocarbons, nitrogen oxides and sulphur dioxide than mineral diesels. Energy security is driven by the volatility of crude oil prices and the perceived threat to national security of over-dependence on foreign supplies. In developing countries, biofuel production can contribute to rural development in three main areas: employment creation, income generation and replacing traditional bio-energy (Brittaine and Lutaladio 2010:6).

In addition, Zoomers (2010) identifies other processes behind the land rush in Africa: (a) the creation of free economic zones and the associated large-scale infrastructure works, normally created in peri-urban zones; (b) large-scale tourist complexes being encouraged in developing countries; (c) rapid increase in 'retirement' (residential) migration, a response to high costs of living in the North, whence many people above 55 are seeking a comfortable existence in cheap sunny environments that have a friendly and caring population; and (d) land purchases by migrants from developing countries (Zoomers 2010:437–40). Smaller and Mann (2009) also note the growing need for land to achieve water security in Africa. Also, the unprecedented economic growth in transition countries has increased the demand for energy. Consumers in these countries have increased their standard of living and biofuel production can damp down the rise in oil prices and thus improve national energy security (Coyle 2007, in Matoni *et al.* 2011:9).

The Actors

Agricultural investment in the past was mainly by Western governments and companies in plantation agriculture for global markets. Now, oil-rich but water-insecure Gulf States like Saudi Arabia, Qatar, and UAE and the emerging giants in Asia like China, South Korea and India increasingly desire to secure land rights and fresh water to supply their domestic food and/or energy needs (Cotula *et al.* 2009; Von Braun and Meinzen-Dick 2009; Smaller and Mann 2009). The scale of foreign investment has increased dramatically in recent years and has generated debate about the benefits and challenges to the livelihoods of the local rural poor. The literature reports that the 'land grab' is taking place either through leasing or selling of the land (Cotula *et al.* 2009; Smaller and Mann 2009; von Braun and Mwinzen-Dick 2009; Friss and Reenberg 2010).

The new investors directly compete with local users of land, threatening their sources of livelihood. These actors include governments of developing countries that are initiating investment, as they are also concerned about the food crisis that rocked the world in 2007-08; financial entities attracted to land-based investments; the greater concentration in agro-processing; and technical advances that favour large operations (Deininger *et al.* 2011:2). These actors are driven by the demand for food and industrial raw materials arising from population and income growth. Demand for biofuels is a reflection of policies and mandates in key consuming countries, including shifts of production of bulk commodities to land-adundant regions where land may be cheaper and the potential for productivity growth higher than in traditional producing regions (Deininger *et al.* 2011:11).

Sub-Saharan Africa is increasingly being targeted as a potential source of agricultural land and natural resources. Foreign governments and private companies are interested in obtaining land to grow crops for food and fuels to meet the growing domestic demand. Biofuel supporters generally argue that biofuel production will address the economic crises of developing countries by creating wealth and jobs and alleviating poverty (Friends of the Earth 2010:4). The region has a comparative advantage in fertile rain-fed soil that is good for biofuel production, as illustrated in Table 1.

Table 1. Availability of agricultural land across regions of the world

Region	Total area (1,000 ha)
sub-Saharan Africa	201, 546
Latin America and the Caribbean	123,342
Eastern European and Central Asia	52, 387
Middle East and North Africa	3,043
Rest of the world	50,971

Source: Fisher and Shah 2010, in Deininger *et al.* 2011. p. 79

Sub-Saharan Africa has the largest area of land suitable for rain-fed agriculture, followed by Latin America and the Caribbean. In these regions, the area of land currently cultivated is large, highlighting the possibly far-reaching social impacts of agricultural investment. Information on land acquisitions is normally at a disaggregated level. It keeps on changing as new deals are negotiated and signed between host and investing governments or private companies. However, some organisations have aggregated the data on land deals, although these vary from source to source, depending on the scope of commercial land deals in a given country. Table 2 illustrates the number of large-scale land acquisitions in select African countries.

Table 2. Large-scale land acquisitions in select African countries

Country	Projects	Area (1,000 ha)	Median size (ha)	Domestic share (proportion of transfers to domestic investors)
Ethiopia	406	1,190	700	49
Liberia	17	1,602	59,374	7
Mozambique	405	2,670	2,225	53
Nigeria	115	793	1,500	97
Sudan	132	3,965	7,980	78

Source: Deininger *et al.* 2011. xxxiii

The statistics on land grabs provided by Friss and Reenberg (2010), after screening and triangulating the scattered quantitative information, reveal that between 51 and 63 million hectares are currently assigned in land deals or under negotiation in 27 African countries. The report also suggests that Madagascar is leading in attracting 16 biofuel projects, that Ethiopia has 15 such projects and that Sudan leads in projects related to food production, mostly from food-insecure Gulf States. It has 11 such projects and Ethiopia eight (Friss and Reenberg 2010:11).

Ethiopia alone has approved 815 foreign-financed agricultural projects since 2007. Land is normally leased to investors for approximately $ 1 per hectare per year. According to the Ethiopian government, it is not the land deals that are causing famine. Rather, these deals have attracted hundreds of millions of dollars in foreign investments and created tens of thousands of jobs. According to one government spokesperson:

> Ethiopia has 74m hectares of fertile land, of which only 15% is currently in use – mainly by subsistence farmers. Of the remaining land, only a small percentage currently in use – 3 to 4% – is offered to foreign investors. Investors are never given land that belongs to Ethiopian farmers. The Government has encouraged Ethiopians in the Diasporas to invest in their homeland. They bring badly needed technology, they offer jobs and training to Ethiopians, they operate in areas where there is suitable land and access to water.[1]

However, social activists in Ethiopia maintain that the land assigned to investors already had owners, who had been using it for centuries, and that land is being given out without consulting the indigenous people, especially in Gambella region and Oromia province.

Sulle and Nelson (2009: 3) report that over four million hectares of land have been requested for biofuel investment in Mozambique, especially for jatropha, sugarcane and palm oil, but only 640,000 hectares have been allocated, and formal rights have been granted to only 100,000 ha. They also report that some land acquisitions for biofuel have targeted land that is used for the forest-based economic activities on which villagers depend heavily (Sulle and Nelson 2009: 4). Mozambique is one of the countries in Africa that promotes large-scale production of biofuels to the extent that the government supports (even financially) their development. Table 3 shows a list of approved biofuel projects according to province.

Sun Biofuels, a UK-based company, operates in both Mozambique and Tanzania. In Mozambique, the project is located in Chimoio in Manica province in the Beira corridor, a primary transport link between the port of Beira and

---

1. http://www.guardian.co.uk/environment/2010/mar/07/food-water-africa-land-grab

Table 3. List of biofuel projects approved by Mozambican government

Province	Biofuel	Area (ha)	Amount in US$
Sofala	ENERTERRA	20, 000 ha	53,305,350
	Crown Energy Zambeze	15,000 ha	224,326,000*
	NIQEL	Not specified	7,500,000
	ZAMCORP-INDICO-CLUSTER AJ1	20,870	12,800
	ELAION AFRICA	Not specified	100,000
Maputo	BIOENERGIA MOCAMBIQUE	6,950 ha	9,600,000
	MOCAMBIQUE INHLAVUKA-BIOCOMBUSTIVEIS	5,348 ha	4,000,000
	ECOMOZ (also to be in Sofala and Nampula)	Not specified	4,000,000
	DEULCO EMVEST	1,220	1,900,000
Inhambane	HENDE WAYELA ENERGIA	Not specified	725,000
	C3 BIO-DIESEL	Not specified	3,000,000
Niassa	LUAMBALA JATROPHA	10,000 ha	400,000
Zambezia	PROJECTO-PILOTO AGRO-INDUSTRIAL DE BIO-COMBUSTIVEIS	1,000 ha	796,500
	QUIFEL AGRICOLA	Not specified	17,535,440
Manica	MOZAMBQUE PRINCIPLE ENERGY	18,000 ha	280,000,000*
	SUN BIOFUELS	5,166,7 ha	7,086,250
Cabo Delgado	OURO VERDE	2,000 ha	730,000
Gaza	ProCAna	30,346 ha	500,000,000*
	ENERGEM-ENRGIAS RENOVAVEIS DE MOZAMBIQUE	Not specified	2, 000 ,000

Source: Mozambican Investment Promotion Centre, April 2011

* Projects that amount to more than US$ 100m are categorised as mega projects

the landlocked countries of Zimbabwe, Zambia and Malawi (see Figure 1). In Tanzania it has acquired 80,000 hectares of land in a degraded natural forest. Sun Biofuels intends to manage at least 20 per cent of the area as conservation and high biodiversity zones, including wetlands and fragile soil areas. It has been operating in Tanzania since early 2006 and was one of the earliest entrants into the biofuels sector in the country. The company's Tanzania operation is located in Kisarawe district. Situated 70 kilometres northwest of Dar es Salaam, at 300 metres elevation and 1,100mm of annual rainfall, it has an ideal location, but one that does not conform to the proposition that jatropha can be grown in very dry lands. Evidence has shown that jatropha does better in fertile than in unfertile soils.

In 2008, Kenya signed an agreement with Qatar for a huge loan to construct a deepwater port in exchange for 40,000 hectares of land in the Tana River area, even though the local community depends on this land as a source of livelihood (Horing 2010:5). Indigenous populations, especially during the dry season, use water from this reservoir for their animals, and for horticulture, fish and subsistence foodcrops.

Figure 1. Sun Biofuels jatropha project in Mozambique

Figure 2. Sun Biofuels 2009[1]

1. http://www.sunbiofuels.com/index.html

Impacts of farmland investment in Africa

Numerous studies have placed biofuel projects high on the development agendas of poor countries. They have been praised because they produce environmentally friendly 'green fuels' (McMichaiel 2010:609), and thus reduce environmental degradation without affecting economic growth (Borras *et al.* 2010:577), a win-win approach. However, biofuels increase pressure on the environment and disadvantage indigenous people and people with insecure land rights (Borras *et al.* 2010:581). 'First generation' biofuel (e.g., palm oil and sugarcane production) would cost poor people a lot in developing countries. This is because biofuel production promotes monoculture, thus adding to the vulnerability of the poor (Dauvergne and Neville 2009 in White and Dasqupta 2010:596).

A study conducted by Cotula *et al.* (2008) found that biofuels can be instrumental in revitalising agricultural land use and livelihoods in rural areas. Small-scale farmers could significantly increase yields and incomes, both of which are necessary for poverty reduction. Large-scale biofuels could also provide employment and encourage skills development and secondary industry. However, these developments are determined by the nature of the land tenure regimes where competing claims exist among local resource users, governments and incoming biofuel producers. Where appropriate conditions are not in place, the rapid spread of commercial biofuel production may result – and is resulting – in poorer groups losing access to the land they depend on and in conflict.

Evidence from Tanzania and Mozambique suggests that biofuel production may offer income-generation opportunities in rural areas and may improve prospects for food security by enabling farmers to purchase food on the market. This represents a new opportunity for farmers in addition to growing traditional crops. However, biofuel production may compete with foodcrops and have significant negative impacts on food security – the so-called food versus fuel debate (Cotula *et al.* 2008:13).

A study by FAO (2010) also found that biofuel offers numerous opportunities to poor countries including increased energy; new markets for producers; employment; poverty reduction and economic growth; and achievement of environmental objectives through reduction of greenhouse gas emissions. However, FAO's study treated these opportunities with great caution, especially their social, economic and environmental viability (FAO 2010:3). Brittaine and Lutaladio (2009) further noted that biofuel production affects water resources and biodiversity, thereby normally leading to declining availability of water for irrigation. Biodiversity is also threatened by large-scale monocropping of exotic species (FAO 2010:9).

According to UNIDO (2010), biofuels are not an environmental panacea and the extent to which they are 'green' or offer carbon savings depends on how they are produced. Biofuel production affects the right to food of millions of people in the medium and long term, especially to groups that need access to fertile soil and clean water to grow their food (UNIDO 2010:10). Although biofuels can play an important role in poverty reduction, they negatively affect vulnerable groups, violating their rights and leading to forced evictions, especially of indigenous peoples, smallholders and forest dwellers, as well as women as land concentration spreads in the rural economy (UNIDO 2010:11).

Nhantumbo and Salomao's study presents Mozambique as having the highest biofuel production potential in Africa. It intends to use biofuels to meet energy demand, create employment and reduce poverty. The study reveals that the 'claim often made that feedstock for biofuels can be commercially grown on marginal land is misleading' (Nhantumbo and Salomao 2010). The ProCana project in Masingir district, Gaza province competes with smallholders for irrigation water from the Limpopo River, leaving little for local farmers (Cotula *et al.* 2008:35). Biofuels production has the potential to compete with production of foodcrops and might reduce access to land by smallholder farmers (Nhantumbo and Salomao 2010:18). The biofuels boom has been associated with tensions between investors and local communities over the acquisition of local land rights and water access for local farmers. In Tanzania, biofuel projects target wetlands (GRAIN 2007; ABN 2007, in Cotula *et al.* 2008:23), and lead to displacement of rural people from their customary lands, as seen in Kisarawe district (African Press Agency 2007, in Cotula *et al.* 2008:37).

SECTION B: Overview of the Dynamics of Property Rights Relations in Africa

Understanding the political dynamics of commercial land deals is very useful in understanding the current land grab in Africa. There are various competing interests in these deals. International institutions such as the World Bank, FAO and IFPRI support foreign investment in farmlands in Africa, along with African governments, which believe that land deals offer a golden opportunity to develop the African agrarian economy.

As noted by Borras and Franco (2010:5), this golden opportunity has been perceived differently, ranging from outright opposition to eager embrace. The dominant social classes and groups and state bureaucrats use their power to lease lands to investors and accommodate corporate interests through expanded food and biofuel production, which swallows up smaller farm units either through purchase or lease. In addition, there is large-scale enclosure of non-private land, especially in Africa, where such land is taken to be marginal (Borras and Franco 2010:22).

Large-scale foreign investment and competition over water and fertile land are likely to fuel more conflicts in the future (Horing 2010). Why? There is hardly any land that is not being used. It is not only the affected community, but also the governments and investors that are at high risk. Investors lack security for their investments, especially in poor countries where there are severe shortages of food, land and water. What is being called idle land and leased by governments to foreign investors in reality often belongs to the community and is governed by customary rights. Idle land is an important source of livelihood for the rural poor, providing them with resources for subsistence farming, such as access to edible wildplants, grazing, water and firewood. The community land resources offer various other benefits as well, including maintaining the water cycle in terms of availability and quality, but have become a target of biofuel and foodcrop production to meet foreign demandsprimarily with the purpose of stabilising food and oil prices in the importing country. Price rises in basic food stuffs such as rice and wheat in 2007–08 encouraged financial speculators to invest in land to generate profit, not to feed people. The fear of climate change poses a major threat to food security, as prices continue to rise due to poor harvests and limited availability of water. For the Gulf States, the major concern is water scarcity. Saudi Arabia's aim in investing in and abroad is to stabilise local food prices and reduce dependency on food imports, as its own groundwater reserves dry up after decades of wheat irrigation.

The recent surges in commercial land deals exhibit broad patterns in changing the features and direction of property rights in a wide range of land policies, including redistribution, distribution, non-(re)distribution and (re)concentration (Borras 2010:17; Borras and Franco 2010:25). Table 4 below illustrates the possible land policy changes.

Table 4. Trajectories of change and reform in land policies

Type of reform	Dynamics of change and reform; flow of wealth and power transfers	Remarks
Redistribution	Land-based wealth and power transfers from landed classes or state or community to landless or near-landless working poor	Reform can occur in private or public lands, can involve transfer of full ownership or not, can be received individually or by group
Distribution	Land-based wealth and power received by landless or near-landless working poor without any landed classes losing in the process; state transfers	Reform usually occurs in public lands, can involve transfer of right to alienate or not, can be received individually or by group
Non-(re)distribution	Land-based wealth and power remain in the hands of the few landed classes or the state or community, that is, status quo that is exclusionary	'No land policy is a policy', also included are land policies that formalise the exclusionary land claims/rights of the landed classes or non-poor elites, including state or community groups
(Re)concentration	Land-based wealth and power transfers from the state, community or small farm holders to landed classes, corporate entities, state or community	Change dynamics can occur in private or public lands, can involve full transfer of full ownership or not, can be received individually, by groups or by corporate entity

Source: Adapted from Borras and Franco 2010. p.17

Within (re)concentration, there are at least three broad trajectories, including:
- Reverse redistribution, where previously redistributed land-based wealth and power (from the landed classes or the state to the working poor) was later redistributed again to the landed classes, other elites or the state.
- Perverse redistribution is where land-based wealth and power are transferred from the working poor to the landed classes, other elites, or the state or the elite community. This includes land reform, forest land allocation or management devolution, formalisation and privatisation of land rights, a variety of land-based joint venture agreements and land lease arrangements, and so on. This might include increased formalisation of land grabbing of indigenous community lands.
- Lopsided distribution is where land-based wealth and power are transferred from the state or community, directly or indirectly, by policy or through the open market, to a handful of private or state entities, with the net effect of excluding others while benefiting a few (Borras and Franco 2010:21–2).

This categorisation by Borras and Franco can help to improve our understanding of the 'tragedy of the commons' in Africa, where communal resources, whether held by the community or by the government on behalf of its people, are being appropriated for commercial agriculture to produce food and biofuels that are not meant for local consumption or to meet local demand, but to meet the interests of the investing countries.

Moreover, unlike the European Union, US, Brazil and Japan, which have well-developed biofuel policies with specific targets (Sieflhorst *et al.* 2008:12), most African governments do not have such policies. Yet, the extent to which a country benefits from biofuel projects depends on policy and institutional environment (Deininger *et al.* 2011:95) mainly with regard to contractual arrangements between investors and local groups, respect for rights of existing users and increasing productivity and welfare in line with existing strategy for economic development. Instead, recipient African governments are faced with a fundamental dilemma: whether to create an enabling and friendly environment for foreign investors or secure the rights of their local populations as well as dealing with new and foreign investors. (Zoomers 2010:443)

As suggested by Smaller and Mann (2009), African governments can develop national policies on biofuel investments based on *domestic international investment contracts* and *international investment agreements*. Under domestic agreements, policy on foreign investors could include admissions, incentives, taxation, land and water rights and environment. International investment contracts explicitly address price, quantity and duration for the purchase or lease of land, taxation and incentives for investors and other operational matters. International investment agreements include bilateral investment treaties, free trade agreements and regional investment treaties, using Most Favoured Nations (MFNs) and New Technology (NTs) principles (Smaller and Mann 2009:9, 11–12).

Smaller and Mann (2009) further address three important questions that have been problematic to many governments in developing countries, such as whether foreign investors have rights to buy land and water rights. What rights do foreign investors acquire if they do invest and what happens to the rights of the previous users of land and water? On the question of whether foreign investors have rights to buy land and water in a host country, the answer is *no*. International law generally does not give investors rights to invest in land and water in another state. However, acquisition of land by foreign investors in another country is fundamentally a matter of domestic law within each state, which may choose to open its economic sector, or not, as it sees fit. What rights do foreign investors acquire if they do invest? In the absence of international contracts or treaties, foreign investors would be treated the same as a domestic investor under the applicable domestic law. However, when a contract between the state and investor exists, the investor may acquire, depending on the terms of contract, additional rights not out in domestic law relating to water use and land tenure rights. What happens to the rights of previous users of land or water, a critical issue in the debate over 'land grab'? Under the domestic law, where the rights are clear and vested in the local owners or users, the later are entitled to be a vendor of the property or water rights, and thus to participate in the contracting process. If government determines that an investment should take place despite the opposition of a landholder, expropriation might be possible, subject to the relevant compensation requirements (Smaller and Mann 2009:14).

Scholars of 'land grab' have frequently pushed for the development of codes of conduct for land acquisitions in Africa. The emphasis has been on ensuring respect by investors of existing land and resource rights, guaranteeing food security and promoting transparency, sharing benefits, environmental stewardship and adherence to national trade policies (Von Braun and Meizen-Dick 2009; Cotula 2009; and Zoomers 2010).

Recently, the World Bank has recognised that large-scale investment poses significant challenges that need to be addressed. In doing so, the World Bank, together with FAO, IFAD, UNCTAD and other development partners have formulated seven principles that all the parties to land deals should adhere to for investment to do no harm, be sustainable and contribute to development. This move is intended to depoliticise commercial land acquisitions in Africa so as to guarantee tenure security for foreign investors. However, landed property is politically embedded by its very nature. That said, these principles must be used by investors and countries involved in large-scale acquisitions.

The seven responsible agro-investment principles are:
1. Respecting land and resource rights;
2. Ensuring food security. Investments do not jeopardise food security, but strengthen it;
3. Ensuring transparency, good governance and a proper enabling environment. Processes for acquiring land must be transparent and monitored, ensuring the accountability of all stakeholders within the proper legal, regulatory and business environment;
4. Consultation and participation. All those materially affected must be consulted, and the agreements from the consultations are recorded and enforced;
5. Responsible agro-investing. Investors ensure that projects respect the rule of the law, reflect industry best practice, are economically viable and result in durable shared value;
6. Social sustainability. Investments generate desirable social and distributional impacts and do not increase vulnerability; and
7. Environmental sustainability. Environmental impacts of a project are quantified and measures are taken to encourage sustainable resource use while minimising and mitigating the risk and magnitude of negative impacts. (Deininger *et al.* 2011:xxvii)

The situation on the ground, however, differs from these policy prescriptions. Biofuel-attracting countries tend to favour foreign investors at the expense of local rights, and existing regulations are rarely followed when sealing land deals with foreign investors. As Horing (2010) suggests, whereas increased investments in agriculture should benefit everybody, there is no responsible investment. The investment should benefit farmers and local populations and improve their livelihoods. It should also respect their rights to land, ecosystems and water resources. Countries must adhere to the principles of responsible investment and the codes of conduct are well known, including: transparency, participation, proper information for the public and people concerned, acknowledgement of existing land and water rights and e proper compensation schemes for those having to be resettled (Horing 2010:5).

The risks arising from new land deals may include neglect of land users, short-term speculation, absence of consultation, corruption, environmental harm, violence and conflicts over land rights, polarisation and instability, undermining food security and loss of livelihood, and failure to keep promises such as local jobs, facilities and compensation (Borras and Franco 2010:8). Hence, there is a need for well-developed land rights; clear identification of land that is available; improved investment climate through rule of law and contract security; evidence-based agricultural policies in relation to incentives, markets, technologies and rural infrastructure; facilitation of contract knowledge and extension services (including rural banking); and decentralised negotiations.

SECTION C: Strategic Choice in Large-Scale Land Acquisitions: Land-Water Access for Commercial Agriculture

The rapid increase in land acquisitions in Africa can be explained by the new hydro-hegemony whereby emerging and developed economies compete for farmlands in Africa. The perspective is that water issues are closely related to agriculture, climate change, economics and politics. Water resources are vital to the success of any proposed investment and considerable thought should be given to water in negotiating investment projects.[2] This article argues that areas in host countries lacking in water resources are not of interest to actors – including foreign governments and investors both foreign and domestic. The water footprint concept is used to describe the demand for blue-water and green-water for commercial agriculture in Africa. The intention is to measure the water dependency in water-scarce countries, including on external water resources.

African wetlands provide great value to local society. They have great cultural and economic importance to indigenous communities. African wetlands cover only about 4 per cent of the total landmass, but store half the world's liquid freshwater. The majority of poor people depend directly on wetlands for their livelihoods due to their high soil fertility.[3] The current literature does not explicitly explain the ownership and control of land and the water nexus. Yet access to water is strategic issue for land being acquired in Africa. Because of the increasing scarcity of water, those who control land will also control water resources. Large-scale land acquisitions in Africa will guarantee access to water that can be used for irrigation purposes. Unfortunately, African governments are negotiating land deals without taking into account the implications for water resources. Moreover, these land deals are rushed through without environmental impact assessments or adequate consideration of the potential implications for neighbouring communities. In fact, in negotiating leases or purchases of African agricultural lands water charges are excluded. The purchase of land by investors in Africa automatically guarantees access to water. This is because Africa's water is held in large rivers, with widespread aquifers, large dams and lakes, and in atmospheric water vapour and soil moisture (UNEP 2010:19). Africa is also blessed with the Nile, the world's longest river. The Nile, as also the Congo and Niger, have dramatic seasonal variability and intra-annual variation that reflects precipitation patterns. Table 5 shows the major African rivers and their key features.

---

2. Achieving water security world-wide, published 10 May 2010, found at http://www.lwrg.org/workshops.html, accessed 3 April 2011

3. www.wetlands.org

Table 5. Major African rivers

River	Basin Area (km²)	Length (km)	Mean Annual Runoff (109 m³)	Unit Runoff	Interesting morphological features
Congo	3,699.1	4,7000	1,260	341	Cataracts at Stanley pool
Nile	3,110	6,850	84	27	Cataracts at Aswan; drains out of large depression – the Sudd
Niger	2,274	4,100	177	78	Has an inland delta; entangled in dune fields
Zambezi	1,388.2	2,650	94	68	Falls at Victoria Falls and Cabora Bassa; linked to the northern Botswana drainage by spillways; entangled in dune fields

Source: UNEP 2010, p. 19

Mountains and other African watersheds contribute greatly to the total stream flow of Africa's major rivers. Such areas mainly receive more rainfall than their lower surroundings. Rivers such as the Nile, the Niger, the Senegal and Orange flow from relatively rain-abundant areas, referred to as the 'water towers of Africa'. These towers are the source of many transboundary rivers, a situation that is sometimes also a source of conflict. Such areas have concentrated, protected and sustainably developed resources to equitably address food security, economic development and environmental issues (UNEP 2010). However, with current large-scale acquisitions in Africa to produce food for foreign markets and biofuels, these poor but resource-rich countries are targeted by rich food-insecure and energy-insecure countries. As shown in Figure 3 and 4 below, the Democratic Republic of Congo (DRC) leads in actual total renewable water resources, followed by Madagascar, Cameroon, then Ethiopia, Mozambique, Sierra Leone, Liberia and Guinea. Therefore, it not surprising that many of these countries are hotspots for large-scale commercial land acquisitions in Africa.

This land-water acquisition nexus indicates that recent large-scale land acquisitions in sub-Saharan Africa are driven by water security for commercial agriculture. Investment in farmlands in Africa is seen as a water insurance policy for protection against increasing food shortages, declining water supplies and climate changes coupled with huge population increases. Whatever the case, one could argue that the acquisition of land and water resources may or may not have a direct link. This article postulates that the acquisition of land in Africa is indirectly related to water acquisition. This is based on trends in investment flows to water-rich countries such as Ethiopia, Sudan, Madagascar, Mozambique, DRC and so forth. Why are investors concentrating their acquisitions in these countries? Land in these countries has high water-resource potential and it is very tempting to invest in crops that require water.

According to a new ranking by Maplecroft (2011), of 186 countries studied the Gulf States are rated as the world's most water-stressed countries, with the least available water per capita. The study was done by calculating the ratio of

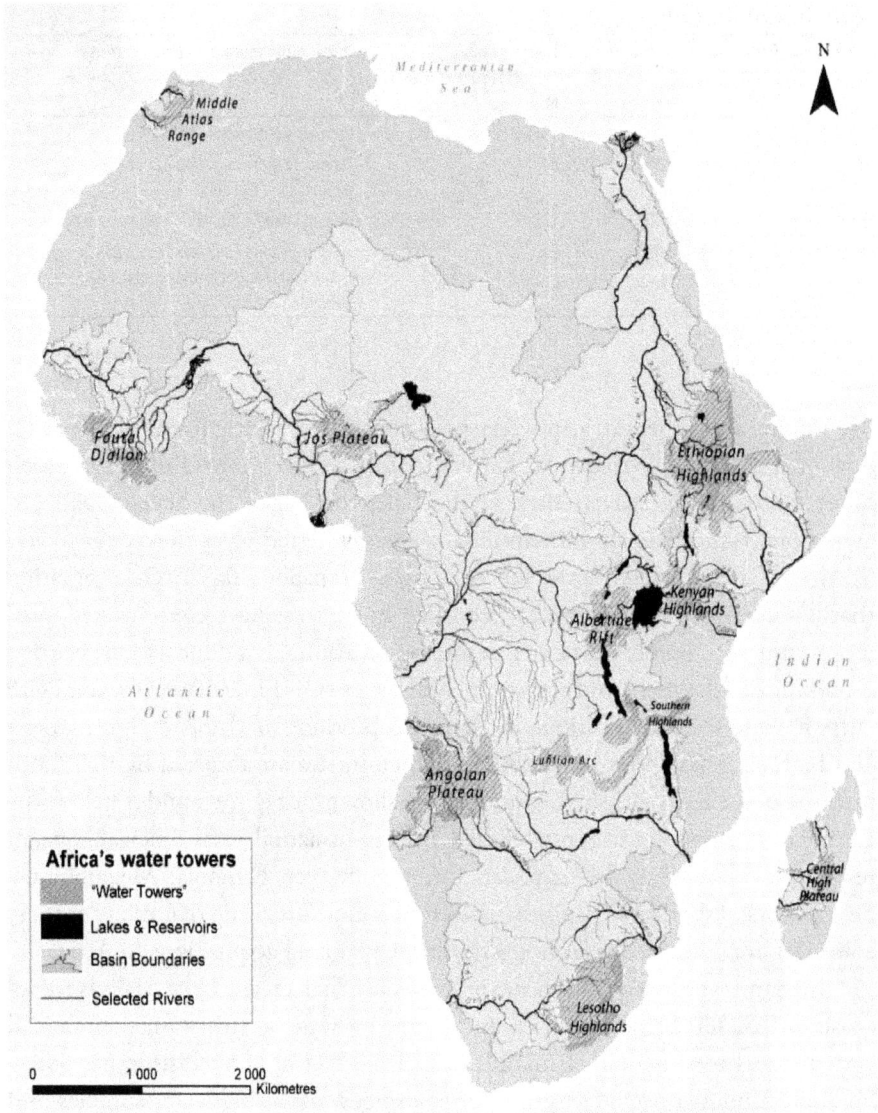

Figure 3. Water towers of Africa
Source: UNEP 2010, pp.6 and 19 respectively

domestic, industrial and agricultural water consumption against the renewable supplies of water for industrial and agricultural water consumption from precipitation, rivers and groundwater. The rating in terms of water stress ranges from extreme risk, high risk, and medium risk to low risk. The Gulf States top the water stress index among Middle Eastern/North African (MENA) countries as illustrated in Table 6 below.

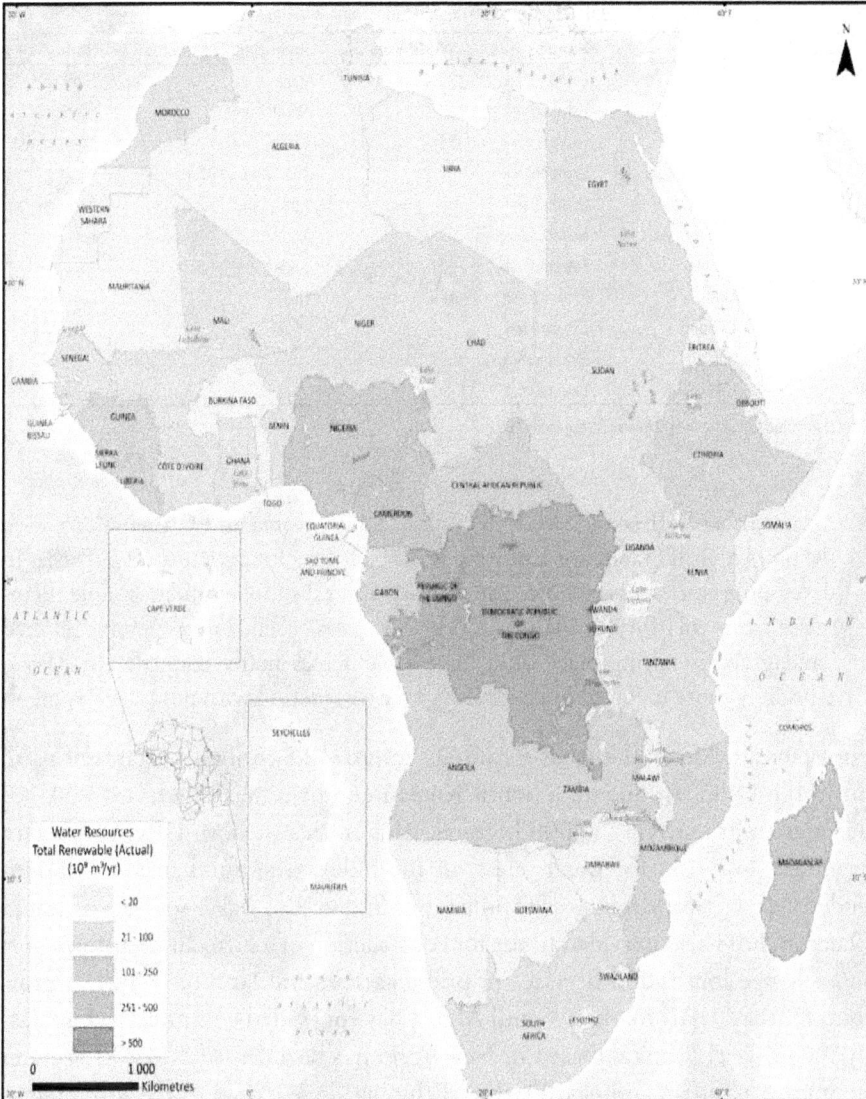

Figure 4. Renewable Water Resources
Source: UNEP 2010, pp.6 and 19 respectively

This Maplecroft study states that the dual drivers of climate change and population growth will combine to squeeze water resources and affect the food security of governments across the world, regardless of how water-secure they may be today. It further points out that water shortages in these countries will be the potential constraint on economic development and could create social unrest if dwindling resources result in higher prices and limited access for their populations. As Maplecroft analyst T. Styles notes:

Table 6. Water stress index for selected MENA countries and emerging economies

World Ranking	MENA	Rating	World Ranking	Emerging Economies	Rating
1	Bahrain	Extreme risk	30	India	High risk
2	Qatar	Extreme risk	36	South Korea	High risk
3	Kuwait	Extreme risk	56	China	Medium risk
4	Saudi Arabia	Extreme risk		sub-Saharan Africa	
5	Libya	Extreme risk	68	Ethiopia	Medium risk
6	Western Sahara	Extreme risk	91	Kenya	Low risk
7	Yemen	Extreme risk	93	Sudan	Low risk
8	Israel	Extreme risk	169	DRC	Low risk
9	Djibouti	Extreme risk			
10	Jordan	Extreme risk			
20	UAE	Extreme risk			

Source: Adopted from Maplecroft Water Stress Index 2011

As a means of offsetting shortfalls, India, South Korea and China, along with the oil rich Gulf States, are acquiring water rich land for agricultural purposes in developing countries to ensure the security of food supplies and decouple themselves from volatility in global food prices. This recent phenomenon, dubbed 'land grab,' is taking place on a huge scale across many countries in Africa, especially those involved in post conflict reconstruction with poor development.

For example, China alone has a contract to grow 2.8 million hectares of palm oil in the DRC. In Sudan, a South Korean company has purchased 700,000 hectares and the UAE 750,000 hectares. This is because both DRC and Sudan are rated 'low risk' for water stress on the index. The Sudd region in South Sudan has the world's largest swamps, but Sudan also has a scarcity of water. Qatar in 2008 also provided funds for construction of a Kenyan deepwater port in exchange for 40,000 hectares of prime agricultural land on which to grow food for the Qatari market.[4] Saudi Arabia has concluded a 42,000 hectare deal in Sudan. Jarch Capital, based in New York, has been leased 800,000 hectares in Southern Sudan, and near Darfur.[5] Ethiopia alone has approved 815 foreign-financed agricultural projects since 2007. Land to investors is normally leased at approximately $ 1 per year per hectare. Ethiopia has 74 million hectares of fertile land, of which only 15 per cent is currently in use – mainly by subsistence farmers. In Oromia province of Ethiopia, where there is water , cheap land and labour and the climate is good, Saudis, Turks, Chinese and Egyptians are looking for land.[6] The government of Mali has granted 100,000 hectares of land in the Macina region to Malibya Company for 50 years to grow hybrid rice for

---

4. http://www.maplecroft.com/about/news/water_stress_index.html, accessed 20 June 2011
5. http://www.guardian.co.uk/environment/2010/mar/07/food-water-africa-land-grab, accessed 20 March 2011
6. Ibid.

export to Libya without the knowledge of local people. The company was also granted priority access to water during the dry season. As a result, local producers' access to water for irrigation from the Niger was reduced significantly (Horing 2010:3). Climate change poses a major threat to food security as prices continue to rise due to poor harvests and the limited availability of water. The major concern is water scarcity. Saudi Arabia's aim in making land investments abroad is to stabilise local food prices and reduce dependency on food imports as groundwater reserves dry up (Horing 2010:4).

Assessing Water Demand of Nations: The Concept of Water Footprint

In order to understand the land-water nexus, this study further explores the water footprint of nations, defined as the total amount of water used to produce goods and services consumed by the inhabitants of each nation. A nation's water footprint is the nation's ability to supply water needed to meet its domestic demands for goods and services (Hoekstra and Chapagain 2008). Water-scarce countries such as Gulf States import virtual water from water-rich countries, thus creating virtual water-import dependency. It is noted that a nation can be water dependent in two different ways: *on water flowing from neighbouring countries* and *on virtual-water imports*. The external water resources of a country constitute a significant part of the total renewable water resources.

Water footprint indicates the volume of water used (cubic metres per year) and comprises blue-water that originates from underground and surface water, and green-water which is infiltrated or harvested water. Grey-water is the polluted ground and surface water. Water footprint only looks at consumption, either from domestic water resources or the use of water outside the country. Water demand in a given country is based on water withdrawals for domestic, agricultural and industrial sectors (Hoekstra and Chapagain 2008:54).

India globally contributes 13 per cent of the water footprint, China 12 per cent, US 9 per cent, Russia 4 per cent, Indonesia 4 per cent and the other countries 58 per cent. The water footprint of a country is determined by a number of factors such as: a) the volume of consumption (related to Gross National Income); b) consumption patterns (either high or low consumption); c) climate and growth (growth conditions); and d) agricultural practices (water efficiency) (Hoekstra and Chapagain 2008:62). Consumption in rich countries contributes relatively more to the water footprint compared to that of poor countries, mainly in the form of consumption of more goods and services, for example meat. The water footprint is greater when consumption of more staple foods such as rice, wheat, maize and meat predominate in consumer behaviour. Also, high consumption of industrial products contributes significantly to the total water footprint of rich countries. Climate conditions, especially in regions where high evaporation makes the water requirement per unit of crop production relatively large, leads to a greater water footprint, for example, the Gulf States. In poor countries, a large water footprint is due to unfavourable climate and bad agricultural practices.

Even though India has the largest water footprint in the world, totalling 987 billion m^3 per year, and also has 17 per cent of total global population, India accounts for only 13 per cent of the global water footprint. Why then should India seek land acquisitions in Africa? The possible reasons are energy and food security for a burgeoning population. As noted earlier, the size of the global water footprint is largely determined by the consumption of food and other agricultural products. The total volume of water used globally for crop production is 6,400 billion m^3 per year at the field level (Hoekstra and Chapagain 2008:55). Crops such as rice take up the largest share of total water volume used for global crop production (up to 21 per cent); followed by wheat (21 per cent); maize (9 per cent); soybean (4 per cent); sugarcane, barley, seed cotton and sorghum (3 per cent); coconut and millet (2 per cent) (Hoekstra and Chapagain 2008:58). Most land acquisitions in Africa are for the purpose of growing foodcrops such as wheat and rice to feed the growing population at home. These crops need a lot of water. Countries that used to import these crops, such as the Gulf States, have instead resorted to growing them, because of the competing use of them as biofuels as oil prices rise. Africa is perceived as the ultimate location for the production of such crops as rice, wheat and maize, which require relatively great amounts of water.

Gulf States have very high degree of virtual import dependency (>50 per cent). Jordan alone annually imports a virtual-water quantity five times its own yearly renewable water resources. In most water-scarce countries, the choice is either between exploiting domestic water resources to increase water self-sufficiency or virtual water imports at the cost of becoming water dependent (Hoekstra and Chapagain 2008:133). The water scarcity of a country is defined as a country's total water footprint divided by the country's water availability. Virtual-water import-dependency is the ratio of the external water footprint of a country to its total water footprint. Countries with a high degree of water scarcity include Kuwait, Qatar, Saudi Arabia, Bahrain, Jordan, Israel, Oman, Lebanon and Malta. Table 7 below illustrates water footprint, scarcity, self-sufficiency and water import-dependency of selected countries.

Fresh water as a geopolitical resource was in the past, and unlike oil, considered a local not a global resource. Water and oil share common economic characteristics, both as factors of production and in being unevenly distributed (Hoekstra and Chapagain 2008:134). As pointed out by various scholars (Donkers 1994, 1999; Barlow and Clarke 2002), water is currently a sort of white oil or blue gold. While virtual water imports have been regarded as a possible cheap alternative source of water in areas where fresh water is relatively

Table 7. Water footprint, scarcity, self-sufficiency and water import-dependency per year, selected countries, 1997–2001

Country	Total renewable water resources (10^9 m³/yr)	Internal water footprint (10^9 m³/yr)	External water footprint (10^9 m³/yr)	Total water footprint (10^9 m³/yr)	Water scarcity (%)	Water self-sufficiency (%)	Water import dependency (%)
China	2.896.57	825.94	57.44	883.39	30	93	7
DRC	1.283	36.42	0.47	36.89	3	99	1
Egypt	58.30	56.37	13.13	69.50	119	81	19
Ethiopia	110.00	42.46	0.42	42.88	39	99	1
India	1,896.66	971.39	15.99	987.38	52	98	2
Jordan	0.88	1.70	4.58	6.27	713	27	73
Kenya	30.20	19.14	2.09	21.23	70	90	10
Korea Republic	69.70	21.02	34.18	55.20	79	38	62
Kuwait	0.02	0.28	1.90	2.18	10.895	13	87
Lebanon	4.41	2.14	4.30	6.44	146	33	67
South Africa	50.00	30.87	8.60	39.47	79	78	22
Sudan	64.50	67.70	0.55	68.25	106	99	1
Libya	0.60	6.77	3.99	10.76	1,793	63	37
Madagascar	337.00	19.51	0.30	19.81	6	98	2
Malawi	17.28	12.87	0.13	13.00	75	99	1
Mali	100.00	21.51	0.13	21.64	22	99	1
Morocco	29.00	37.02	6.58	43.60	150	85	15
Mozambique	216.11	19.43	0.05	19.49	9	100	15
Oman	0.99	0.91	2.92	3.83	389	24	76
Qatar	0.05	0.19	0.43	0.62	1,176	31	69
Tanzania	91.00	36.53	0.99	37.51	41	97	3
Yemen	4.10	6.86	3.84	10.70	261	64	36
Zambia	105.20	7.27	0.25	7.52	7	97	3

Source: Adopted from Hoekstra and Chapagain 2008, Globalization of Water: Sharing the Planet's Freshwater Resources

scarce to ease pressures on domestic water supplies, governments in water-scarce countries currently want to buy land in Africa. These countries used to import water-intensive products from countries rich in water resources. However, the current oil crisis has changed this practice. Water-scarce countries such as the Gulf States are looking at farmlands in Africa as an alternative to imports. Food crops such as wheat that used to be imported cheaply have become expensive because of their competing use in biofuel production.

Imports of virtual water – as opposed to real water, which is generally too expensive – relieves the pressure on nations if products are traded between countries with high water productivity to countries with low water productivity. Because resources are unevenly distributed between the haves and have-nots, and scarcity is fuelled by current climate change, the increasing dependency of water-scarce nations on the supply of water footprint can be exploited for the benefit of nations (Hoekstra and Chapagain 2008:132). China and India have the highest degree of national water self-sufficiency (93 per cent and 98 per cent respectively). China has a relatively low water footprint per capita, 700 m³ per year compared to India's 980 m³ per year. However, changes in consumption patterns in these countries have resulted in water scarcity, raising the question of how China and India will sustain themselves in the future. If they want to increase food security through food imports, they may need land and water resources in the rest of the world (Hoekstra and Chapagain 2008:134).

Biofuel Projects and Land-Water Access

It is noted here that biofuel projects target African water resources as opposed to the belief that biofuel crops are grown on marginal African lands. Freshwater has become a global resource because of trade in water-intensive commodities. Demand for freshwater comes from the production of first generation biofuels. The European Union target to replace 10 per cent of transport fuels with renewables by 2020 requires additional water for the production of first generation biofuels. In addition, the global water footprint for biofuel-based transport will be 9 per cent of current global water footprint. Use of first generation biofuels increases competition for freshwater resources unless nations adopt more water-efficient second generation biofuels (Gerbens-Leenes and Hoekstra 2011:2658). This highlights the fact that freshwater availability has become a challenge and that increases in agricultural water use intensify competition for water. As noted by Gerbens-Leenes (2011:2663), the expansion of crop production for biofuels leads to a large increase in freshwater use coupled with an increase of water stress in some countries. They further assumed that the increase in water demand is in proportion to the increase in biofuels demand (Gerbens-Leenes and Hoekstra 2011:2665). On sustainability criteria, for example in the European Union, the law protects forests and greenhouse sequestration. The water footprint is unsustainable when the process is located in a 'hotspot' where for a certain period of the year total water use is unsustainable. When more water is withdrawn, nature is affected. In fact, while the European Union sustainability criteria focus on the protection of untouched nature and on greenhouse gas savings, first generation biofuels in Africa are produced through deforestation of pristine forests and or recently degraded areas, drained peatland, wetlands or in areas of high biodiversity.

Africa's most common biofuel crops include sugarcane, maize, sweet sorghum and cassava for bioethanol, while palm oil and jatropha are common sources of biodiesel (Sieflhorst et al. 2008:19). Most biofuel crops require a lot of water for their survival, and this explains why they compete with local water uses and why most projects are concentrated in water-rich areas like swamps and forest reserves, as illustrated in Table 8 below.

Table 8. Biofuel crops and their level of need for water resources

Biofuel crop	Wetland conservation	Irrigation	Fertilisers and pesticide use	Sensitivity to water supply	Preferential rainfall (mm/year)	Ethanol/ Oil Yields (l/ha)	Required economic scale for competitive biofuel production (ha)
Ethanol							
Sugarcane	High	High	High	High	1,500 -2,500	4.00 – 8.00	17,500
Maize	Low	Medium to high	High	High	700 -1,500	700 – 3,000	N/A
Sweet sorghum	Low	Medium	High	Low to medium	400 - 650	3,000 – 6,000	15,000
Cassava	Low	Low	Medium	Low to Medium	1,000 – 1,500	1,750 – 5,400	15,000
Biodiesel							
Palm oil	High	Low	Low	High	1,800 – 5,000	2,500 – 6,000	400 – 4,000
Jatropha	Low	Medium	Medium	Low to Medium	600 – 1,200	400 – 2,200	400- 1,000

Source: Sieflhorst *et al.* 2008, pp. 27, 28 and 30

Based on Table 8 above, both sugarcane and palm oil plantations need a lot of water for good yields and have better energy yields than jatropha. Their production has negatively reduced wetland areas, with resulting declines in water quantity and quality. Sugarcane yields well in tropical wetland areas. It also requires large amounts of water throughout the year as well as much land for its production. This could explain why most bioethanol projects in Africa are being acquired in areas with abundant water. Unfortunately, in most cases the land tenure systems in these areas are informal, with the land being owned by the community, which derives a livelihood from it. Examples from Tanzania, Kenya, Mozambique and Uganda show the considerable expansion of biofuel plantations into wetlands.

In Uganda, the BIDCO project is a greenfield palm oil plantation in Kalangala district. It is located on one of the islands of Lake Victoria. The area allocated to it was part of the forest reserve. The government of Uganda in 2004 approved this project and BIDCO company was allocated 10,000 hectares of tropical rain forest. While the company was investing in the project, the government offered a corporate tax holiday and VAT deferral as part of the agreement with the company. This project was strongly opposed by environmentalists and members of parliament because of the ecological impact and government's payment of import excise duties on behalf of BIDCO. In northern Uganda, the Madhvani group of companies has since 2006 been negotiating for communal land in Amuru district for use as a sugarcane plantation. The first target was to get 40,000 hectares of land along a stretch of the Nile used as communal property and as a source of livelihood (such as farming, food gathering, hunting, firewood), and many clans in Acholi claim ownership of the land. Land is thus regarded as a common heritage. Customary land is owned by individual holders only for purposes such as cropping and grazing, but selling this land is subject to family approval. The clan or family can prohibit the sale of clan land to 'undesirable persons' and any transaction can be declared void if it does not receive approval. The community in general has the right to graze communally over the whole area, but not to damage crops, and enjoys the right of free access to water points for animals (such as rivers and springs) and to other common resources.

Most of the biofuels projects in Mozambique have been allocated in areas with high water potential. The ProCana Project has been given 30,000 hectares of land in Masingir district, Gaza province to grow sugarcane. The project is supposed to draw water from a dam fed by a tributary of the Limpopo, a dam that also supports irrigated smallholder agriculture. Farmers downstream have expressed concern that the project will deplete the available water, leaving little for them (Cotula et al. 2008:35).

In addition, a study in Mozambique reveals that the 'claim often made that feedstock for biofuels can be commercially grown on marginal land is misleading'. One company has switched from jatropha to forestry due to poor soils. Fertile lands and water are necessary for commercially grown biofuels (Nhantumbo and Salomao 2010:4). Biofuel crops such as sugarcane, sweet sorghum and even jatropha require soil with reasonable fertility and access to water. Most existing and planned sugarcane projects are in areas with easy and abundant access to water. In Tanzania, 400,000 hectares of land in Wami Basin have been targeted by a Swedish investor for sugarcane production (GRAIN 2007; ABN 2007, in Cotula et al. 2008). Currently, jatropha farming is in Mbamba Bay on Lake Nyasa, and in the coast and Lake Victoria regions.[7]

---

7. Tanzania: Government's Serious Leadership Needed in Regulating Biofuels, http://www. ngonewsafrica.org/?p=1136, accessed 20 March 2011

SECTION D: Common Resources and Water Access Management in Africa

This section links water rights to land rights as they are intertwined in customary arrangements. The fact that land that is supposedly 'idle', 'marginal' or 'abandoned' is in places where water resources are concentrated, for example swamps and forested areas, should not obscure the fact that traditionally such communal land was used during dry seasons. No individual is allowed to claim ownership of these communal resources because everybody is given access. Large-scale land acquisitions always target these communal resources in Africa. The point of contention centres on ownership, possession and use. It is helpful to understand what constitutes common property rights. Who controls what? And how is access to land and water resources managed? In this way one can better understand differences between ownership rights and possession rights. In customary arrangements, ownership rights exist without possession rights. It is the social group that is considered the owner of the land. This can be a clan, a kinship group or a family. Every member of the social group had the right of ownership and had an obligation to see that this right was maintained and observed. In some instances, individuals were allowed to transfer land temporarily, but no monetary consideration was involved. Hence, individuals only exercise 'possession rights' while 'ownership rights' remain communal. Common property is the right of a group of individuals not to be excluded from its use or the benefits that can be derived from it (Alden-Willy 2011; Cockburn 2002).

Variations in power (i.e., rights) are derived from social relations, not the market. Rights over land are trans-generational and control is exercised through members of the units of production and is not simply the product of 'political superordination'. Different land uses attract degrees of control at different levels of sociopolitical organisation: allocations of arable land are, for instance, often controlled at family level, while grazing involves a wider community (Okoth-Ogendo 1989:11). Rights of individuals and families vary from discrete temporary uses, such as gathering natural resources in a communal forest, grazing on communal pastures, cultivating a specific field for one or several seasons, to permanent control over a piece of land or other resource for cultivation and bequest to heirs (Lastarria-Cornhiel 2002).

In some rural societies, sales of land are forbidden by law and custom. Either the law simply does not allow it or does so in the most complicated or difficult way, and customs are interpreted by chiefs and may even be used to prevent people from asserting their claims as owners (Lund 2002:7). The occupation and use of a piece of land is the main evidence of ownership or an existing interest in the land. Access to land is contingent upon tribal or community membership controlled by the chief. There are strong and exclusive residential rights to grazing land and natural resources (Mzumara 2003:2).

While foreign large-scale land acquisitions in Africa may be driven by the need for green fuels, Gulf States are driven by food production needs, and all the while the dual factors of climate change and population growth squeeze water resources in the region. The assumption that the land-water nexus holds value is evidenced in the general discussion. Gulf States and emerging economies like India, South Korea and China are mostly targeting water-rich countries such as Sudan, Ethiopia and the DRC. They are increasingly concerned about the availability of freshwater, which is becoming a scarce commodity in home countries, especially in the Gulf States, and in the populous emerging economies. Sub-Saharan Africa is now seen as an insurance policy for water security due to the perception that it has huge water potential, a favourable climate, plentiful land and relatively cheap local labour.

Although virtual water imports are regarded as a cheap alternative source of water in water-scarce countries that can ease pressure on domestic water supplies, Gulf States are currently buying land in Africa to produce food. These countries used to import water-intensive products from water-rich countries. The current oil crisis has changed this practice. Gulf States are looking at farmlands in Africa as an alternative. Food crops such as wheat that used to be imported cheaply have become expensive because of their competing use in biofuel production.

The perception that land is abundant in Africa is misleading. The nature of land tenure and property rights in Africa is complicated. In most cases, land is already used or claimed under customary regimes: it is only that people do not have formal land rights. Customary land rights are associated with common resources such as wetlands and forests and foreign investors are interested in land with greater irrigation potential, which are more likely to be already in use by indigenous people. Indigenous 'moral property' rights have been preserved under customary laws for common resources such forests, rangelands, hunting areas and water resources. Land in Africa is normally left vacant for various reasons, including environmental stewardship, such as fallowing to prevent soil erosion. Land can also be used for livestock grazing, as communal hunting grounds and as a source of medicine and fuels. The issue of unused land does not generally apply in African contexts. Individualisation in these areas is not sanctioned. Because customary tenure has not been recognised explicitly in most countries' land laws, and in some cases is not recognised at all, it is at the mercy of the state, which has vested interests on behalf of the people. In fact, the large-scale land acquisitions in Africa are communal land acquisitions in which states that should have been the trustee on behalf of the community gives the lands to foreign investors. These acquisitions have led to the displacement of people without adequate consultation and compensation, to broken promises and to opportunities to violate indigenous rights to land, food sovereignty and water access. These large-scale land acquisitions have not been matched by invest-ment commitments in the deals and enforceability of commitments by recipient countries remains weak, especially in countries that have just emerged from conflict. Hence, there must be a stabilisation of land rights before further in-vestment takes place to protect the rights of existing land users – the indigenous people, women, pastoralists whose rights are held under customary tenure.

References

Alden-Willy, L., 2011, *The Tragedy of Public Lands: The Fate of the Commons Under global commercial pressure*. Rome: International Land Coalition.

Barlow, M. and T. Clarke, 2002, *Blue gold: The battle against corporate theft of the world's water*. New York: New Press.

Bernstein, H., 2010, *Class Dynamics of Agrarian Change*. Halifax: Fernwood.

Borras, S.M. Jr., P. Michael and I. Scones, 2010, "The Politics of biofuels, land and agrarian change: Editors' introduction", *Journal of Peasant Studies*, Vol. 37, No. 4, October, pp. 575-92.

Borras, S.M. Jr. and J. Franco, 2010, *Towards a Broader Views of Politics of Land Grab: Rethinking Land Issues, Reframing Resistance*. ICAS Working Paper Series No. 001.

Brittaine, R. and N. Lutaladio, 2010, *Jatropha: A small holder bioenergy crop. The potential for pro-poor development*. Integrated Crop Management, Vol. 8. Rome: FAO.

Cockburn, C.J., 2002, *Property and Credit: Property Formulation in Peru*. Cambridge, MA: Lincoln Institute of Land Policy Working Paper.

Cotula, L., N. Dyer and S. Vermeulen, 2008, *Fuelling Exclusion? The biofuels boom and poor people's access to land*. Rome: FAO and IIED.

Cotula, L., S. Vermeulen, R. Leonard and J. Keeley, 2009, *Land grab or development opportunity? Agricultural investment and international land deals in Africa*. London/ Rome: FAO, IIED and IFAD.

Daniel, S. and A. Mittal, 2009, *The Great Land Grab: Rush for World's Farmland Threatens Food Security for the Poor*. Oakland, CA: Oakland Institute.

Dauvergne, P. and J.K. Neville, 2010, "Forests, food, and fuel in the tropics: The uneven social and ecological consequences of the emerging political economy of biofuels", *Journal of Peasant Studies*, Vol. 37, No. 4, October, pp. 631–60.

Deininger, K., D. Byerlee, J. Lindsay, J. Norton, A. Selod and M. Stickler, 2011, *Rising Global Interest in Farmland in Africa: Can It Yield Sustainability and Equitable Benefits?* Washington DC: World Bank.

Donkers, H., 1994, *De witte olie: Water, vrede en duurzame ontwikkeling in het Midden-Oosten*. Utrecht: Uitgeverij Jan van Arkel.

FAO, 2010, *Bioenergy and Food Security. The BEFS Analytical Framework*. Rome: FAO.

Friends of the Earth International, 2010, *Africa: up for grabs. The scale and impact of land grabbing for agrofuels*. Ugbowo, Nigeria: Environmental Rights Action,/Friends of Earth Nigeria.

Friss, C. and A. Reenberg, 2010, *Land grab in Africa: Emerging land system drivers in a teleconnected world*. Global Land Project Report No. 1, GLP International Project Office, University of Copenhagen.

Gerbens-Leenes, W. and A.Y. Hoekstra, 2011, "The water footprint of biofuel-based transport", *Energy Envision. Science*, Vol. 4, pp. 2658–68.

GRAIN, 2008, *The 2008 landgrab for food and financial security.* Barcelona: GRAIN.

Höring, U., 2011, *Water and Land Grabbing.* Geneva: Ecumenical Water Network and Ecumenical Advocacy Alliance.

Hoekstra, A.Y. and A.K. Chapagain, 2008, *Globalization of Water: Sharing the Planet's Freshwater Resources.* Oxford: Blackwell.

Lastarria-Cornhiel, S., 1997, "Impact of privatization on gender and property rights in Africa", *World Development,* Vol. 25. No. 8, pp. 1317–33.

Lund, C., 2002, "Negotiating Property Institutions: The symbiosis of property and authority in Africa", in Juul, K. and C. Lund (eds), *Negotiating property in Africa,* pp. 144-62. Uppsala: Nordiska Afrikainstitutet.

Matondi, P.B., K. Havnevik and A. Beyene, 2011, "Introduction: Biofuels, food security, and land grabbing in Africa", in Matondi, P.B., K. Havnevik and A. Beyene (eds), *Biofuels, land grabbing and food security in Africa.* London: ZED Books.

McMichael, P., 2010, "Agrofuels in the food regime", *Journal of Peasant Studies,* Vol. 37, No. 4, October, pp. 609-29.

Mzumara, D,. 2003, *Land Tenure Systems and Sustainable Development in Southern Africa.* ECA/SA/EGM/Land/2003/2. Lusaka: Economic Commission for Africa.

Nhantumbo, I. and A. Salomao, 2010, *Biofuels, land access and rural livelihoods in Mozambique.* London/Maputo: International Institute for Environment and Development and CentroTerra Viva.

Okoth-Ogendo, H.W.O., 1989, "Some Issues of Theory in the Study of Tenure Relations in Africa Agriculture", *Africa,* Vol. 59, No. 1, pp. 6–17.

Roundtable on Sustainable Biofuels (RSB), 2011, *RSB Land Rights Guide-Lines.*

Sieflhorst, S., M.J. Molenaar and D. Ofermans, 2008, *Biofuels in Africa: An Assessment of Risks and Benefits for African Wetlands.* Amsterdam and Wageningen: AIDE Environment and Wetlands International.

Smaller, C. and H. Mann, 2009, *A Thirst for Distant Lands: Foreign investment in agricultural land and water.* Winnipeg: International Institute for Sustainable Development.

Sulle, E. and F. Nelson, 2009, *Biofuels, and Land Access and Rural Livelihoods in Tanzania.* London: IIED.

Von Braun, J. and R. Meinzen-Dick, 2009, "Land Grabbing by Foreign Investors in Developing Countries: Risks and Opportunities", *IFPRI Policy Brief 13.* Washington: International Food Policy Research Institute.

United Nations Environment Programme, 2010, *Africa: Water Atlas.* Nairobi.

UNIDO, 2010, "Making it Industry for Development: Wind of Change?" *Quarterly Magazine,* Number 2, Vienna International Centre.

White, B. and A. Dasgupta, 2010, "Agrofuels capitalism: A view from political economy", *Journal of the Peasant Studies,* Vol. 37, No. 4, October, pp. 593-607.

Zoomers, A., 2010, "Globalization and the foreignation of space: Seven processes
    driving the current global land grab", *Journal of the Peasant Studies,* Vol. 37, No. 2,
    April 2010, pp. 429–47.

CURRENT AFRICAN ISSUES PUBLISHED BY THE INSTITUTE

Recent issues in the series are available electronically
for download free of charge www.nai.uu.se

1. *South Africa, the West and the Frontline States. Report from a Seminar.* 1981, 34 pp, (out-of print)

2. Maja Naur, *Social and Organisational Change in Libya.* 1982, 33 pp, (out-of print)

3. *Peasants and Agricultural Production in Africa. A Nordic Research Seminar. Follow-up Reports and Discussions.* 1981, 34 pp, (out-of-print)

4. Ray Bush & S. Kibble, *Destabilisation in Southern Africa, an Overview.* 1985, 48 pp, (out-of print)

5. Bertil Egerö, *Mozambique and the Southern African Struggle for Liberation.* 1985, 29 pp, (out-of print)

6. Carol B.Thompson, *Regional Economic Polic under Crisis Condition. Southern African Development.* 1986, 34 pp, (out-of-print)

7. Inge Tvedten, *The War in Angola, Internal Conditions for Peace and Recovery.* 1989, 14 pp, (out-of print)

8. Patrick Wilmot, *Nigeria's Southern Africa Policy 1960–1988.* 1989, 15 pp, (out-of-print)

9. Jonathan Baker, *Perestroika for Ethiopia: In Search of the End of the Rainbow?* 1990, 21 pp, (out-of-print)

10. Horace Campbell, *The Siege of Cuito Cuanavale.* 1990, 35 pp, (out-of-print)

11. Maria Bongartz, *The Civil War in Somalia. Its genesis and dynamics.* 1991, 26 pp, (out-of-print)

12. Shadrack B.O. Gutto, *Human and People's Rights in Africa. Myths, Realities and Prospects.* 1991, 26 pp, (out-of-print)

13. Said Chikhi, Algeria. *From Mass Rebellion to Workers' Protest.* 1991, 23 pp, (out-of-print)

14. Bertil Odén, *Namibia's Economic Links to South Africa.* 1991, 43 pp, (out-of-print)

15. Cervenka Zdenek, *African National Congress Meets Eastern Europe. A Dialogue on Common Experiences.* 1992, 49 pp, ISBN 91-7106-337-4, (out-of print)

16. Diallo Garba, *Mauritania– The Other Apartheid?* 1993, 75 pp, ISBN 91-7106-339-0, (out-of print)

17. Zdenek Cervenka and Colin Legum, *Can National Dialogue Break the Power of Terror in Burundi?* 1994, 30 pp, ISBN 91-7106-353-6, (out-of print)

18. Erik Nordberg and Uno Winblad, *Urban Environmental Health and Hygiene in Sub- Saharan Africa.* 1994, 26 pp, ISBN 91-7106-364-1, (out-of print)

19. Chris Dunton and Mai Palmberg, *Human Rights and Homosexuality in Southern Africa.* 1996, 48 pp, ISBN 91-7106-402-8, (out-of print)

20. Georges Nzongola-Ntalaja *From Zaire to the Democratic Republic of the Congo.* 1998, 18 pp, ISBN 91-7106-424-9, (out-of print)

21. Filip Reyntjens, *Talking or Fighting? Political Evolution in Rwanda and Burundi, 1998–1999.* 1999, 27 pp, ISBN 91-7106-454-0, SEK 80.-

22. Herbert Weiss, *War and Peace in the Democratic Republic of the Congo.* 1999, 28 pp, ISBN 91-7106-458-3, SEK 80,-

23. Filip Reyntjens, *Small States in an Unstable Region – Rwanda and Burundi, 1999–2000,* 2000, 24 pp, ISBN 91-7106-463-X, (out-of print)

24. Filip Reyntjens, *Again at the Crossroads: Rwanda and Burundi, 2000–2001.* 2001, 25 pp, ISBN 91-7106-483-4, (out-of print)

25. Henning Melber, *The New African Initiative and the African Union. A Preliminary Assessment and Documentation.* 2001, 36 pp, ISBN 91-7106-486-9, (out-of print)

26. Dahilon Yassin Mohamoda, *Nile Basin Cooperation. A Review of the Literature.* 2003, 39 pp, ISBN 91-7106-512-1, SEK 90,-

27. Henning Melber (ed.), *Media, Public Discourse and Political Contestation in Zimbabwe.* 2004, 39 pp, ISBN 91-7106-534-2, SEK 90,-

28. Georges Nzongola-Ntalaja, *From Zaire to the Democratic Republic of the Congo. Second and Revised Edition.* 2004, 23 pp, ISBN-91-7106-538-5, (out-of print)

29. Henning Melber (ed.), *Trade, Development, Cooperation – What Future for Africa?* 2005, 44 pp, ISBN 91-7106-544-X, SEK 90,-

30. Kaniye S.A. Ebeku, *The Succession of Faure Gnassingbe to the Togolese Presidency – An International Law Perspective.* 2005, 32 pp, ISBN 91-7106-554-7, SEK 90,-

31. Jeffrey V. Lazarus, Catrine Christiansen, Lise Rosendal Østergaard, Lisa Ann Richey, *Models for Life – Advancing antiretroviral therapy in sub-Saharan Africa.* 2005, 33 pp, ISBN 91-7106-556-3, SEK 90,-

32. Charles Manga Fombad and Zein Kebonang, *AU, NEPAD and the APRM – Democratisation Efforts Explored.* Edited by Henning Melber. 2006, 56 pp, ISBN 91-7106-569-5, SEK 90,-

33. Pedro Pinto Leite, Claes Olsson, Magnus Schöldtz, Toby Shelley, Pål Wrange, Hans Corell and Karin Scheele, *The Western Sahara Conflict – The Role of Natural Resources in Decolonization.* Edited by Claes Olsson. 2006, 32 pp, ISBN 91-7106-571-7, SEK 90,-

34. Jassey, Katja and Stella Nyanzi, *How to Be a "Proper" Woman in the Times of HIV and AIDS.* 2007, 35 pp, ISBN 91-7106-574-1, SEK 90,-

35. Lee, Margaret, Henning Melber, Sanusha Naidu and Ian Taylor, *China in Africa.* Compiled by Henning Melber. 2007, 47 pp, ISBN 978-91-7106-589-6, SEK 90,-

36. Nathaniel King, *Conflict as Integration. Youth Aspiration to Personhood in the Teleology of Sierra Leone's 'Senseless War'.* 2007, 32 pp, ISBN 978-91-7106-604-6, SEK 90,-

37. Aderanti Adepoju, *Migration in sub-Saharan Africa.* 2008. 70 pp, ISBN 978-91-7106-620-6, SEK 110,-

38. Bo Malmberg, *Demography and the development potential of sub-Saharan Africa.* 2008, 39 pp, 978-91-7106-621-3

39. Johan Holmberg, *Natural resources in sub-Saharan Africa: Assets and vulnerabilities.* 2008, 52 pp, 978-91-7106-624-4

40. Arne Bigsten and Dick Durevall, *The African economy and its role in the world economy.* 2008, 66 pp, 978-91-7106-625-1

41. Fantu Cheru, *Africa's development in the 21st century: Reshaping the research agenda.* 2008, 47 pp, 978-91-7106-628-2

42. Dan Kuwali, Persuasive Prevention. *Towards a Principle for Implementing Article 4(h) and R2P by the African Union.* 2009. 70 pp. ISBN 978-91-7106-650-3

43. Daniel Volman, *China, India, Russia and the United States. The Scramble for African Oil and the Militarization of the Continent.* 2009. 24 pp. ISBN 978-91-7106-658-9

44. Mats Hårsmar, *Understanding Poverty in Africa? A Navigation through Disputed Concepts, Data and Terrains.* 2010. 54 pp. ISBN 978-91-7106-668-8

45. Sam Maghimbi, Razack B. Lokina and Mathew A. Senga, *The Agrarian Question in Tanzania? A State of the Art Paper.* 2011. 67 pp. ISBN 978-91-7106-684-8

46. William Minter, *African Migration, Global Inequalities, and Human Rights. Connecting the Dots.* 2011. 95 pp. ISBN 978-91-7106-692-3

47. Musa Abutudu and Dauda Garuba, *Natural Resource Governance and Eiti Implementation in Nigeria.* 2011. 74 pp. ISBN 978-91-7106-708-1

48. Ilda Lindell, *Transnational Activism Networks and Gendered Gatekeeping. Negotiating Gender in an African Association of Informal Workers.*
2011. 44 pp. ISBN 978-91-7106-712-8

49. Terje Oestigaard, *Water Scarcity and Food Security along the Nile. Politics population increase and climate change.*
2012. 92 pp. ISBN 978-91-7106-722-7

50. David Ross Olanya, *From Global Land Grabbing for Biofuels to Acquisitions of African Water for Commercial Agriculture.*
2012. 41 pp. ISBN 978-91-7106-729-6

www.ingramcontent.com/pod-product-compliance
Lightning Source LLC
Chambersburg PA
CBHW080058280326
41934CB00014B/3354